POLLUTION

POLLUTION

GERALDINE AND HAROLD WOODS

A GROLIER COMPANY
FRANKLIN WATTS
NEW YORK | LONDON | TORONTO | SYDNEY | 1985
A FIRST BOOK

Photographs courtesy of:
AP/Wide World: pp. 2, 13;
Electric Power Research Institute: p. 7;
EPA/Steve Delaney: pp. 8, 17, 31, 38, 47;
Dr. Roger J. Cheng/State University
of New York at Albany: p. 18, 54;
Esso Research and Engineering Company: p. 23;
U.S. Air Force: p. 30;
USDA-Soil Conservation Service: p. 41;
UPI/Bettmann Archive: p. 42;
U.S. Department of Energy: p. 48.

Library of Congress Cataloging in Publication Data

Woods, Geraldine.
Pollution.

(A First book)
Bibliography: p.
Includes index.
Summary: Describes the causes of pollution, air and water pollution, pesticide pollution, garbage and toxic wastes, and the dangers of nuclear pollution.
1. Pollution—United States—Juvenile literature.
[1. Pollution] I. Woods, Harold. II. Title.
TD180.W66 1985 363.7′3 84-20982
ISBN 0-531-04916-7

Printed in the United States of America
5 4 3 2 1

CONTENTS

POLLUTION

INTRODUCTION

A TALE OF TWO APRILS

April 22, 1970—Earth Day. All across America, hundreds of thousands of people staged festivals, parades, and demonstrations in support of a cleaner earth. In Washington, D.C., girl scouts in canoes fished trash from the Potomac River. High school students in Omaha, Nebraska, collected 156,000 used cans for recycling. Young people in West Virginia picked up five tons (4.5 mt) of garbage from a local highway.

Environmentalists (people who are concerned about the physical well-being of the planet) had high hopes on Earth Day. Americans were beginning to understand that our air, water, and soil had been *polluted*, or dirtied, with all kinds of chemicals and poisons. It seemed as if the time had come when we could really begin to clean up the *environment*.

In the years following Earth Day, that's exactly what happened. Many laws were passed, and a government department called the Environmental Protection Agency was created to fight pollution. Private citizens and many businesses looked for ways to help. Much progress was made—but not enough.

On April 22, 1970, thousands of people congregated in Philadelphia's Fairmount Park to observe Earth Day and express their support for a cleaner environment.

April 22, 1980—the tenth anniversary of Earth Day. A tremendous explosion ripped through Elizabeth, New Jersey. A thick, black plume of smoke laced with orange flames rose into the sky. An old dump, containing 20,000 or more drums of dangerous chemicals, had caught fire. It took more than ten hours to control the blaze. Schools were closed, and children were told to stay in their houses because of the smoke and poisonous gases. As people prepared to celebrate a clean environment once again, ash from the Elizabeth fire drifted through the air.

On the first Earth Day, a young girl in New York carried a sign saying, "We treat this world of ours as if we had a spare one in our back pocket."

This book was written because we don't.

CHAPTER ONE

THE WEB OF LIFE

Before the next century is over, a few daring men and women will probably enter a gleaming new spaceship for a flight to Mars or beyond. The journey will last several years. While in space, the astronauts will be completely cut off from Earth. Therefore, the ship will have to be stocked with enough food, water, air, and fuel to last for the entire trip. An accident to any part of the ship's basic supplies—a leak of a poisonous chemical into the food storage area, for example—could end in tragedy for everyone on the spaceship.

Environmentalists often compare our planet to a spaceship. We are, at every moment, traveling through space at tremendous speed. We are also cut off from any new supplies of the basic needs of life. All the water that ever existed and ever will exist is now on earth. So is all the air. It is very unlikely that we will ever harvest food from any planet but our own.

NATURAL CYCLES

Luckily for us, "Spaceship Earth" was designed to renew itself. Water, for example, is constantly falling as rain and collecting in lakes, rivers, and other bodies of water. Some water seeps through soil to underground deposits; some is used by people, plants, and animals. Water is returned to the clouds by evaporation (a change into a gas) where it is available to form rain and begin the cycle again. Food is also cycled and recycled or used over again. Plants take nourishment from the soil and energy from the sun. When they die, plants decay or are eaten by animals, who themselves eventually die and decay. In either case, the plant material is returned to the soil where it nourishes new growth.

Of course, not all of our "spaceship" supplies are renewed so easily. Fuels such as oil and natural gas were formed over the course of millions of years by natural forces within the earth. They are now used at a far faster rate than they are replaced.

In order to be sure that "Spaceship Earth" does not run out of the basic elements of life, we must conserve or recycle resources and energy. We must also protect what we have from pollution. Since the earth's population is increasing every year, we just do not have supplies to spare.

THE WEB OF LIFE

Curbing pollution is very difficult. Every day patches of land are paved over, new crops are grown, and factories are built. All of this affects the earth's air, water, and soil, as well as the animals that live on the planet. This is because all life on earth is bound together in *ecosystems*—the life forms and natural forces in a particular area that affect each other. A change in one part of an ecosystem may change all the other parts. About a hundred years ago, for example, farmers in Jamaica were losing about a fifth of their sugar cane crops to rats. One farmer brought mon-

Fish that swim in polluted waters ingest poisons that are passed on to humans in even greater concentrations when humans eat the fish.

gooses from India to control the rats. The mongooses did eat the rats, but they also ate birds, crabs, and lizards—animals that normally eat insects. With fewer enemies, the insects multiplied rapidly. They ate much more sugar cane than the rats ever had!

THE FOOD CHAIN

Pollution can have far-reaching effects as it moves through an ecosystem. Suppose a small amount of *pesticide* spills into a stream. The water might contain 1/100 parts per million of the poison. That is, for every million cups of water there is 1/100 cup of pesticide. It's a small amount, but the tiny plants that live in the stream absorb the water, and concentrate the poison in their bodies at a strength of 3 parts per million (a million cups of plants and 3 cups of poison). Fish eat the plants, and soon contain 300 parts per million of the poison in their bodies. Animals eat the fish, and end up with a poison concentration of 3000 parts per million. As the poison travels up the food chain, it becomes more and more dangerous.

Since we cannot get off our "spaceship" easily, it's important to understand the pollution problems that face us today—what causes them, what is being done now to clean up the earth, and what might be done in the future.

CHAPTER TWO

AIR POLLUTION

In Los Angeles, children are asked to play indoors three or four days a week. Little League games are canceled, and high school football practices are moved into the gym. These activities aren't rained out—they are "aired" out. That's because the air in Los Angeles often becomes too dangerous to breathe. Comics like to joke that natives of the city don't trust air that they cannot see or chew.

A BOTTOMLESS GARBAGE PAIL

However, air pollution—in Los Angeles and elsewhere—is no laughing matter. There are 5,700,000,000,000,000 tons (554,040,000,000,000 mt) of air, or atmosphere, blanketing the earth. With that much air around, it's not surprising that for many years people treated the atmosphere like a bottomless garbage pail. However, smoke, dust, poisonous gases, and other pollutants don't really go away. They stay in the air, or settle into soil or water. Far too often, pollutants find their way into human lungs, where they can cause serious diseases.

Many of the pollutants in the air come from natural sources. Volcanoes and forest fires produce smoke and soot; the wind whips dust from the ground. Tiny life forms in soil, called bacteria, produce the gas *nitrogen oxide*. Other gases, called *hydrocarbons*, are released when plants and animals decay. They all go into the air. Plants also send small grains of pollen into the air.

Added to these natural pollutants are those from man-made sources. Wherever coal, oil, and natural gas are burned, nitrogen oxide and a gas called *sulphur dioxide* pour into the air. American factories, power plants, cars, and home heating systems produce 130,000,000 tons (117,936,000 mt) of sulphur dioxide and 53,000,000 tons (48,081,600 mt) of nitrogen oxide each year. These gases make life miserable for city dwellers. Sulphur and nitrogen oxides irritate people's eyes and lungs, destroy the metal frames of cars, and rot the stone on city buildings.

Burning fuel also releases carbon dioxide gas into the air. Carbon dioxide is a natural ingredient in our atmosphere. In the past century, the amount of carbon dioxide in the atmosphere has increased by 10 percent. By the year 2000, the increase may be as high as 25 percent. Since carbon dioxide traps heat from the sun, some scientists are afraid the added gas may permanently change the earth's climate. A warmer climate might melt some of the polar ice, raise ocean levels, and affect farming all over the world. Industry also sends 3,000,000 tons (2,721,600 mt) of poisonous metals such as arsenic, cadmium, mercury, nickel,

On August 7, 1980, in New York City, a tanker truck began leaking its 9,000 gallon cargo of explosive propane gas, forcing the closing of the George Washington Bridge and creating a massive traffic jam.

pick 6

and zinc into the atmosphere. When they are breathed in, these metals can cause cancer and other diseases.

POLLUTION MACHINES

Los Angeles's problem stems partly from its geography. The ocean on one side of the city and the mountains on the other create a "trap" for pollutants so they can't escape. The pollutants enter the air from factories and also from the mighty freeways that encircle the city. Anyone who has sniffed the air in the center of a traffic jam knows that cars are efficient pollution machines. Cars puff nitrogen oxide and hydrocarbons from their exhaust pipes. These two pollutants can combine in the presence of ultraviolet light from the sun to form another gas called *ozone*. Ozone often mixes with other pollutants to form *smog*. Smog, which gets its name from the words "smoke" and "fog," is a gray mass of polluted air that stings eyes and makes breathing difficult. It's during smog alerts that Los Angeles's children and old people are asked to stay indoors.

Perhaps the most dangerous automobile pollutant is the invisible, odorless gas *carbon monoxide*. Eighty-seven million tons (78,926,400 mt) of this gas are released into the air each year in the United States. Small amounts of carbon monoxide can cause headaches; large amounts can kill people. Many older cars also give off lead, which can build up in the human bloodstream and cause brain damage.

CLEAN AIR OR AT LEAST CLEANER

Los Angeles began its efforts to control air pollution in the 1940s, and is still following a detailed plan that should end smog alerts by the year 2000. America as a whole began to clean the atmosphere above it in 1970, when Congress passed the Clean Air Act. The Clean Air Act strictly regulated the amount of pollution that could be sent into the air by factories and cars. Many other laws to decrease air pollution have also been passed since 1970. For example, cars built since 1975 have attachments that break down some of the dangerous exhausts. New cars also burn lead-free gasoline.

These pollution control laws have improved America's air. The amount of lead in the air decreased by 55 percent between 1976 and 1980. In twenty-two of the most polluted cities, the number of "smog alerts" dropped 36 percent in the past ten years. Sulphur dioxide in the air was cut by 5,000,000 tons (4,536,000 mt) a year since 1975.

However, there is still a long way to go. In New York City, there are unhealthy levels of pollution in the air over a hundred days a year. As of January 1983, 111 counties in America had air that was too dirty to meet government standards.

OZONE

The air surrounding us is not the only part of the atmosphere causing concern. There is a natural layer of the gas ozone high in the atmosphere. Though large amounts of ozone near the surface of the earth are harmful, the high ozone layer blocks dangerous rays from the sun. If it were to disappear, there would be as many as 300,000 more cases of skin cancer each year in the United States. That's why many people are concerned about spray cans of deodorant, paint, whipped cream, and so forth. Every time a button is pushed on one of those cans, a tiny bit of gas called *fluorocarbon* is released into the air. The fluorocarbon drifts hundreds of miles into space, where it removes some ozone from the planet's protective layer. For this reason many people are switching to "pump" style containers, which contain no fluorocarbons.

POISON FROM AFAR

Scientists have recently become aware that air pollution can have deadly effects hundreds of miles from its source. As many as 50,000 lakes and streams in the United States and Canada are dead or dying because of acid-rain pollution. Acid rain is caused by sulphur and nitrogen oxides that are caught in high wind currents and blown hundreds of miles away. The pollutants drift with the clouds until rain washes them from the air. When combined with rain, the gases change into another form. Sulphur dioxide becomes sulphuric acid, and nitrogen oxide becomes nitric acid. The acid rain falls into lakes and streams. In some places, natural substances absorb the acids, but in other places, the pollutants build up. The water becomes as sour as lemon juice.

This has a disastrous effect on life. Fish cannot produce any young in acid water. So as the fish in the lake die off, they are not replaced. Underwater plants die, and so do the insects that eat the plants. The birds that feed on fish and insects are the next to go. Soon the lake is clear and quiet—as quiet as a grave.

Besides killing fish, acid rain also affects trees and plants on land. There are a few crops that do better with acid rain, such as corn, strawberries, and tomatoes. However, many others are harmed. When soybeans are watered with acid rain, the size of the seeds and the number of pods on the plants are cut almost in half. In the Green Mountains of Vermont, some scientists think that as much as thirty percent of the forests have died in the past twenty years because of acid rain.

High smokestacks emit sulphur and nitrogen oxides into the atmosphere, eventually resulting in acid rain which falls on faraway trees, lakes, and streams.

This photograph shows the damage that acid rain can cause to a single leaf.

Acid rain can also draw metals out of the soil and pipes it touches. Fish can absorb the metals and pass them along the food chain. If the water is used for drinking, the dissolved metals enter human bodies where they cause serious diseases.

EXPENSIVE SOLUTIONS

There are several ways to reduce the damage done by acid rain. Limestone can be added to lakes and streams to absorb the acid. Low sulphur coal, or coal that has been washed before burning, can be used for fuel. "Scrubbers," devices that clean smoke with water before it comes out of smokestacks, cut down on pollutants. Scientists are also studying ways of burning limestone with coal, to trap pollutants before they escape into the air.

Unfortunately, all these solutions are expensive. Some power companies, which burn much of the coal used in America, think that their customers' electric bills might rise as much as fifty percent if pollution controls are tightened. (However, other groups believe that prices for electricity would only rise four or five percent.) Many people in industry also claim that not enough is understood about the causes and cures of acid-rain pollution to make an expensive clean-up program worthwhile. They point out that sulphur and nitrogen oxides can come from natural sources, such as volcanoes.

Pollution control has a high price tag, but so does acid rain, which causes up to five billion dollars worth of damage each year. Moreover, in 1983, the National Academy of Sciences stated firmly that most acid rain does come from man-made sources. The Academy also said that the problem must be solved quickly, before permanent damage to the environment becomes too great. Many scientists and environmentalists have recommended that the sulphur dioxide now being released into the air be immediately cut by at least fifty percent.

CHAPTER THREE

PESTICIDE POLLUTION

In 1955, roofs fell in all over Borneo. The deadly disease malaria almost disappeared from the island, and cats began to fall from the sky. All these incidents were the side effects of a poison that was used to kill germ-carrying mosquitoes.

The poison killed more than mosquitoes. Flies and cockroaches also died; so did the lizards that ate them. Next, the cats that ate the lizards died. Without cats, the rat population grew, and there was an increase in diseases caused by rat fleas. Lastly, wasps and other insects that eat caterpillars died. Soon there were huge numbers of caterpillars on Borneo. They all munched happily on the islanders' straw roofs, which then collapsed. To help the islanders, the British Air Force parachuted bundles of cats all over Borneo. The new cats brought the rats and fleas under control fairly quickly, though it took much longer for the rest of the wildlife to come back into balance.

PESTICIDES IN THE FOOD CHAIN

The poison used on Borneo was DDT, a strong pesticide. Pesticides are chemicals that kill pests like mice, rats, and mosquitoes, as well as nuisance plants like crab grass and poison ivy. Pesticides are not new inventions; the Chinese used dried flower petals to control pests more than 2000 years ago. Until this century, most pesticides were made from natural substances like tobacco leaves. These natural substances often lost their poisoning power fairly quickly. In the last forty years, however, many chemical pesticides have been invented. The new pesticides are usually stronger than the natural products, and sometimes remain poisonous for a longer period of time.

A long-lasting pesticide is especially dangerous because it has a chance to build up in the food chain. Scientists estimate that only 1 percent of the total amount of pesticide used reaches its target. The remaining 99 percent affects other animals and the environment. The pesticide may also pass through the food chain until it reaches human beings. Because pesticides have washed into drinking water supplies and have been sprayed on most food crops, human beings regularly eat small amounts of pesticides. Recent tests show that almost every American has small amounts of at least one type of pesticide in his or her body.

A THREAT TO THE ENVIRONMENT

The lizards and cats on Borneo died because they ate too much DDT. DDT and a pesticide named *Endrin*, which were used in the 1950s and 1960s to control insects that attack cotton, killed thousands of fish in the Gulf of Mexico. Many birds died after eating those fish; others laid eggs with extremely thin shells. As the eggs broke, the chicks inside died. The brown pelican, the state bird of Louisiana, disappeared from that state completely because of DDT and Endrin. Pesticides also upset the balance of nature, as in Borneo, allowing some types of animals to multiply too rapidly.

Another problem with pesticides is tolerance. When a pesticide is first used, many insects die. Those that survive are able to resist its poisonous effect. The surviving insects breed and pass

Pesticides are sprayed on a Texas citrus grove.

their resistance to their offspring. The pesticide becomes less and less effective. In the early 1960s, for example, DDT was used to control malarial mosquitoes in the Asian country Sri Lanka. By 1966, there were only seventeen cases of malaria there. By 1970, however, the mosquitoes that survived were tolerant of DDT, and there were 4,000,000 cases of malaria in Sri Lanka. Scientists estimate that there are now about 400 pests that can tolerate one or more pesticides. To solve this problem, stronger pesticides or new ones with unknown effects are produced. These pesticides may have an even greater effect on the environment.

A THREAT TO HEALTH

Each year about 200 Americans die and 14,000 more become ill from the effects of pesticides. Though pesticides are strictly regulated by the government, accidents often expose many more people to low doses of the poisons. In 1982, for example, the pesticide *heptachlor* was found in milk on the island of Hawaii. Heptachlor had been sprayed on pineapple plants which were later fed to dairy cattle. The pesticide passed into the cows' milk, and was drunk by people throughout the state before it was discovered.

No one really knows the long-term effects of pesticides in the human body. Since diseases like cancer can take twenty to thirty years to develop in humans, scientists usually test the chemicals on smaller animals with shorter lifespans. If the animals develop cancer, the chemical may be a threat to human beings. Each year, as science advances, the tests become more sensitive, and dangers that were previously unknown are discovered. So some pesticides are used for several years before their risks are known. In this way, large quantities of deadly poisons can build up in the environment.

That's what happened with DDT. It was used in the United States for many years before studies showed that it could cause cancer in laboratory animals. There was also evidence that DDT had built up in the food chain. It was even found in the milk of many nursing mothers. Because of the health risks, the United States outlawed the use and sale of DDT in 1972, though the chemical is still used in some other countries.

A WEAPON AGAINST STARVATION AND DISEASE

In spite of all these problems, pesticides have been very useful. Over a billion and a half pounds (680 million kg) of other chemical pesticides are used each year in the United States alone. Many more tons are applied around the world. Yellow fever, bubonic plague, typhus, and sleeping sickness are just a few of the deadly diseases pesticides have helped to control. Pesticides are also useful in agriculture. The United Nations estimates that each year insects and weeds destroy enough crops to feed a billion people. A quarter of all the people in the world who starve to death do so because of losses due to pests. If no pesticides were used in America, crop production would drop by 30 percent. Prices for the remaining food would rise as much as 50 percent. Though no one likes to use poison, not many people are in favor of disease and starvation.

However, environmentalists point out that there are other methods of controlling pests. Insects, for example, are often used to get rid of other insects! Many farmers buy containers of ladybugs and set them loose in their fields. The ladybugs eat aphids—tiny bugs that destroy tomatoes and other plants. Ladybugs are also used to control a disease that attacks pine trees. Another good method is to treat insects so that they are unable to have babies. When the treated insects are set free, they mate with healthy partners. Since no young are produced, the number of pests drops. Unfortunately, this method can backfire. In 1981, thousands of fruit flies were treated and set free in California. However, the treatment was not done properly. The fruit flies produced many babies, and threatened crops all over the state. Scientists have also developed some new crops that pests tend to avoid. Those crops need much less pesticide than other varieties do.

CHAPTER FOUR

GARBAGE AND TOXIC WASTE

Two or three cars, 10,000 bottles, 27,000 bottle caps, 17,500 cans, and 252,000 pounds (114,307 kg) of other garbage—that's what the average American throws away during his or her lifetime. Where does all that trash go? Unfortunately, too much of it ends up in parks, on the roadside, and in America's waterways. Nobody likes to look at litter, but there seem to be many people who don't mind scattering it behind them as they travel through life. Unless it is picked up, litter can pollute the landscape for a very long time. An aluminum soda can takes 500 years to decay. A plastic bag lasts 240 years; a glass bottle may remain undisturbed for as long as a million years.

Even garbage that is properly thrown away can end up as an environmental problem. Some towns burn their trash, sending smoke and soot into the air. Others place it in *landfills*, low-lying or swampy areas that are gradually "filled in" with garbage. When they are well designed, landfills are clean and safe. However, landfills must be vented properly, because decaying garbage gives off an explosive gas called methane. In one landfill, seeping methane caught fire and killed two workers. Liquid also oozes from the bottom of landfills as more garbage is piled on top. The liquid, which is called *leachate*, must be collected and purified so that it doesn't flow through the ground and mix with drinking water supplies.

A good way to start solving the garbage problem is to simply produce less garbage. Forty percent of America's trash comes from packaging material—boxes, containers, fillers, and so forth. Environmentalists would like to see groceries and other goods enclosed in less wrapping. Some scientists have taken this idea to the extreme; they've developed food packaging that can be eaten! The meal *and its box* are placed in boiling water, where the packaging dissolves and becomes part of the meal.

Another way to get rid of trash is to recycle it. The earth has a limited amount of resources. It doesn't make sense to bury or burn things that could be used again. Though much more is needed, some materials are already being recycled: the gray lining inside a cereal box was probably a newspaper at one time. Many aluminum cans, and some bottles, are melted down and made into new containers. To encourage recycling, some states, such as Oregon and New York, require a small deposit for soft drink and beer bottles and cans. When the container is returned to the store, the deposit is refunded. Some children earn extra money by picking up bottles that have been left as litter and bringing them in for the deposit.

Garbage can make money in other ways. Saugus, Massachusetts, burns garbage from Boston and its suburbs, and sells the steam to a power company. Many communities collect methane from their landfills and sell it as fuel. Completed landfills also create new space for people's use. Mountain View, California, for example, is filling in a flood plain. When it is finished, the "dump" will be planted with grass and turned into a golf course.

DANGEROUS WASTES

Garbage can be smelly, ugly, dirty, and occasionally deadly. In Niagara Falls, New York, for example, the Hooker Chemical

Company dumped its wastes into an unused trench called Love Canal for thirteen years. The wastes were a poisonous stew. Many of the chemicals could cause cancer and other illnesses, or damage unborn children. In 1953, Hooker covered the canal with clay and earth, and gave the land to the city of Niagara Falls. Hooker explained that no houses should be built on the canal, because of the wastes underneath it.

The city constructed several roads across the canal. Along its edges, houses and a school were built. A few years later, strange black liquid began to seep into the basements of homes in the area. Some children received burns while playing in the playground. In many basements near the canal, the air smelled foul. People in the area became sick with various diseases, and many pregnant women lost their unborn babies.

After a long investigation, scientists found that the wastes had escaped from the canal and flowed through the ground to the houses nearby. Eight hundred families were forced to move; their neighborhood became a ghost town, surrounded by a tall fence.

Love Canal isn't the only environmental disaster caused by *toxic wastes.* In the early 1970s, oil was sprayed on dirt roads in Times Beach, Missouri, to settle dust. The oil contained small amounts of *dioxin.* Dioxin is an impurity that is produced in the manufacture of many chemicals, such as a pesticide called "2,4,5-T" or "Agent Orange." Agent Orange kills plants; it was sprayed on Asian jungles from 1962 to 1971 during the Vietnam War. Though scientists know of no human beings who have died directly from the effects of dioxin, they still consider it an extremely dangerous substance. It can cause skin irritation, liver disease, and cancer. Many veterans of the Vietnam War are now suing the maker of Agent Orange, claiming that their health was ruined by the chemical.

In Missouri, the dioxin stayed undisturbed for twelve years. Then in the winter of 1982, heavy rains washed the poison into the town. Again, families had to leave their homes and again, a neighborhood became a ghost town.

Above: *the U.S. Air Force spraying defoliation chemicals over South Vietnam in 1967.* Opposite: *chemical wastes stored in metal drums eventually corrode the containers and escape, polluting the water and soil.*

A HUGE PROBLEM

Over 200 million tons (181,440,000 mt) of toxic waste is produced in the United States each year. Before 1976, when Congress passed a strict law regulating toxic waste, poisonous chemicals were often left in garbage dumps or in metal drums. Now, some toxic wastes are injected into deep wells, usually in rock far inside the earth. Others are placed in landfills with thick clay or plastic linings through which wastes cannot pass. The leachate that seeps from the landfill must be collected in ponds and treated to remove the poisons. This is done so that none of the toxic wastes mixes with drinking water supplies.

Unfortunately, these safety measures are not always enough to prevent accidents. Clay or plastic linings can crack, allowing leachate to drain into the earth. Deep wells can explode; once wastes injected into a well in Canada blew out of the ground in Michigan!

Even after 1976, many tons of toxic chemicals have been dumped illegally. The law says that dangerous chemicals must be kept track of throughout the manufacturing process, from the first moment they are used to the moment they are thrown away. However, this law is sometimes ignored. One businessman in New Jersey filed papers saying that he had placed 270,000 gallons (103,000 dl) of waste in a landfill in Delaware. Investigators found that the "landfill" was really an office building. No one knows where the wastes really went. They may have been taken away by "midnight dumpers"—criminals who take truckloads of poisonous liquid and dribble them along country roads, drop them in sewers, or spray them on fields. Each year, as much as 40 percent of America's toxic waste is illegally dumped.

Some toxic wastes are legally recycled. For example, almost all of the alcohol used to wash yarn in the clothing industry is now sold to other companies. In San Francisco, "pickle acid," a waste product from the manufacture of steel, is sold to a power plant north of the city. Some toxic wastes may also be mixed with oil and used in power plants. However, the oil must be burned at an extremely high temperature in order to destroy the dangerous chemicals.

Scientists are now studying new methods of treating toxic wastes, such as concentrating the poisons into a thick paste, or mixing them with cement. Solid wastes are safer than liquids, since they are not able to seep or flow away. Tiny bacteria may also help solve the toxic waste problem someday. Researchers have found bacteria that will eat some types of chemical wastes and change them into harmless substances. They are now searching for a life form that can gobble dioxin.

Today, the Environmental Protection Agency estimates that there are eighty thousand dumps in the United States that contain toxic wastes. Over a thousand of those may endanger public health. In 1980, the government taxed oil and chemical companies to create a "Superfund." Superfund pays for the cleanup of the worst of the dumps, whenever the person or company responsible is unknown or not available. However, Superfund money is enough for only one hundred and seventy of the thousand dangerous dumps, and it will be years before even that work is done.

CHAPTER FIVE

WATER POLLUTION

One day in 1969, a fireboat in Cleveland received an urgent call. The boat was ordered to sail as quickly as possible to the Cuyahoga River. When the fireboat arrived on the scene, the crew found no ship or pleasure boat in flames. Instead, the river itself was on fire. The fireboat had to "put out" the burning water!

The Cuyahoga in those days was a smelly stream of brown liquid, covered with a haze of oil. Greaseballs 18 inches (45.7 cm) wide floated on its surface. There were almost no fish swimming in the murky water; birds and other wildlife also stayed away. Today, the Cuyahoga no longer stinks, and the grease and oil are gone. Fish and birds have returned, though the water is still not totally clean.

FACTORIES WHOSE PRODUCT IS POLLUTION

The Cuyahoga became polluted and then partially recovered in much the same way as countless other bodies of water all across

America. The factories that line its banks use enormous quantities of water in the manufacturing process. It takes 100,000 gallons (37,850 dl) of water to make a single car, and 200 more gallons (75.7 dl) of water to process enough gas to fill a gas tank. Even something as small as a gold wedding band requires 2500 gallons (946.3 dl) of water for its manufacture.

Factories return most of the water to its source after it has been used. However, the water is not always clean when it pours back into its original river or lake. At times it is mixed with dangerous chemicals. A General Electric plant in New York, for example, dumped about 25 pounds (11.3 kg) of chemicals called *polychlorinated biphenyls*, or *PCB*s into the Hudson River every day for many years. PCBs were once used to make paint, ink, plastic, and electrical products, before anyone realized they were dangerous. When studies showed that PCBs could cause cancer in laboratory animals, the dumping stopped and PCBs were no longer manufactured. But by that time, 440,000 pounds (1,995,000,000 kg) of PCBs were on the bottom of the Hudson. PCBs were also deposited in many other bodies of water all over America. The chemicals were found in fish, and had worked their way up the food chain to human beings. In 1975, one-third of all nursing mothers tested had PCBs in their milk.

Though PCBs have been banned, many other chemicals still pour into America's rivers and lakes every day. The Ohio River, for example, has 300 to 600 different types of chemicals in it at all times. Since many of America's rivers and lakes are used for drinking water, chemical pollution often ends up flowing from household faucets. When the tap water from seventy-nine United States cities was tested, at least six chemical pollutants were found in every city. No one knows the long-term effects of this kind of pollution.

Industry also pollutes with heat; water that returns from a factory is often much warmer than the natural temperature of the river or lake. Some types of fish and plants die because of the extra heat, though others do well. However, if the plant is closed for repairs and the temperature drops suddenly, fish that depended on the warmth also die.

It's obvious that the balance of nature is very delicate; even substances we don't usually think of as harmful can damage the environment. In 1981, for example, a brewery in Olympia, Washington, poured 500,000 gallons (189,250 dl) of beer that didn't meet its quality standards into Capitol Lake. A week later, 50,000 fish were dead.

DOWN THE DRAIN

Factories aren't the only source of water pollution. Almost every time a toilet flushes, water and waste material enter the sewer system. (Houses that are not connected to sewer systems usually have buried tanks to collect wastes.) The sewer system leads to a sewage treatment plant where water is purified before it is released into a river, lake, or ocean.

Unfortunately, the sewage treatment plants of many American towns can't handle all the wastes they receive. Some were built many years ago, when fewer people lived in the area. When the sewage overflows, "raw" sewage or dirty water is dumped into the environment. Since human wastes often carry germs that cause serious diseases, the bodies of water they are dumped into must be closed to fishing and swimming. Divers who are forced to swim in polluted waters are often ill for weeks.

Water that flows through sink drains also enters the sewage system. Floor cleaners, shampoos, waxes, and other household chemicals flow down these drains. Yet most sewage systems are only equipped to handle germs, and cannot remove chemicals from the water. So these pollutants enter the nation's water supply. Many streams in the United States have soap bubbles floating on them, left over from detergent that once cleaned someone's clothes. Detergents that contain chemicals called *phos-*

phates are especially harmful because they encourage the growth of the tiny plants *algae*. Although each individual plant is small, millions and millions of them can fill up a body of water and destroy the other life in it.

RUNOFF

What's on city streets? According to one study in Milwaukee, Wisconsin, the city's roads were coated with 5400 pounds (2449 kg) of lead, 7600 pounds (3447 kg) of zinc, 2100 pounds (953 kg) of copper, 170 pounds (77 kg) of chromium, and 12 pounds (5.4 kg) of PCBs. Salt also covers our roads; 9,000,000 tons (8,164,800 mt) of salt are spread on American streets and highways each year to melt snow and ice. That's one-tenth of the entire world's salt production! All of these substances can threaten human health if they are taken into the body in large enough amounts. Many can also harm plants, and rot underground wires and pipes. Since not many people eat off the pavement, this kind of pollution might not seem dangerous. However, rain washes pollutants from the street into sewers, where it "runs off" into the water supply.

Runoff is also a problem outside the city. Fertilizers and pesticides that are sprayed on lawns and fields can wash into rivers and streams. The chemicals can also soak through dirt until they reach underground water deposits. The same thing can happen to toxic wastes that are improperly buried. This is an urgent problem, since half of the people in the United States drink water from wells drilled into these underground deposits, and well pollution can last for hundreds of thousands of years. According to a 1978 government report, forty-two states have seriously polluted

A layer of harmful soap covers much of this pond.

groundwater. Twenty-nine percent of America's cities with more than 10,000 people have polluted wells. In one town in New Jersey, 148 wells have been capped since 1974 because poisonous chemicals were found in the water.

NINE DROPS TO DRINK

Three-quarters of our planet is water; it seems strange to worry about running out of the precious liquid. Yet if all the water on earth were equal to ten gallons (3.7 dl), only *nine drops* would be drinkable. The rest is salty, frozen in ice caps and glaciers, or too deep in the earth to be available to human beings.

Many parts of the United States have already had water shortages during periods of light rainfall. Some areas, such as the Southwest and the Great Plains, may not be able to supply enough water for their residents as early as 1999. So it makes sense to keep the water we have pure and drinkable. To achieve this, the government passed the Federal Water Pollution Control Act, also known as the Clean Water Act, in 1972. According to this law, the country's rivers and lakes were to be made "fishable and swimmable" by 1983. Thirty-seven billion dollars were spent to stop new pollution and clean up what was already in the water supply. Local governments also made water-pollution-control efforts. In many areas detergents containing phosphates were outlawed. Factories were ordered to clean their waste water before dumping it back into the environment. New sewage treatment plants were built.

The deadline for clean water has now passed, and there have been many successes. The Cuyahoga is cleaner. Lake Erie, which was once declared "dead" by scientists, has become clean enough to allow some of its beaches to reopen. However, many American lakes and rivers are still not "fishable and swimmable." Boston, for example, still pours 12 billion gallons (45,420,000 dl) of raw sewage into its waters each year. Fish are now returning to the Detroit River; only a few years ago ducks

Manure and nutrient-laden runoff from this farm's feedlot have washed into an otherwise clear flowing stream.

died simply because they landed on its surface. Scientists estimate that 65 miles (104 km) of Chesapeake Bay "dies" every summer because of pollution from sewage plants and chemicals.

A new Clean Water Act is now being considered. It may include controls on runoff, which was not regulated under the old law. It may also classify lakes and rivers according to use; bodies of water that are used only for boating may be allowed more pollution than those that are used for drinking. Of course, the best solution would be to make all of America's water completely pure. Unfortunately, it will cost over two billion dollars just to reach the lower standards by the year 2000.

THE MIGHTY OCEANS

Some people consider the ocean, like the atmosphere, big enough to absorb any amount of pollution. Many tons of industrial wastes are dumped beneath the waves each year. Until 1970, America placed dangerous nuclear wastes into drums covered with concrete, and left them on the sea floor. Some European countries continue to do this. Raw sewage is also poured into the ocean in many areas. New York City, for example, dumps 200 million gallons (75,700,000 dl) of untreated sewage into the Atlantic every day, though this practice will stop when two new treatment plants are completed in 1987.

Accidents add to the problem. Up to 240,000,000 gallons (912 mil l) of oil is shipped to the United States every day. Even more oil is produced in offshore wells drilled into the sea floor. Each year, over 10,000,000 gallons (38 mil l) spill into American waters

Two Eastham, Massachusetts, residents hold ducks covered with oil from an offshore oil spill.

from wells or the giant tankers that carry the oil. Oil can kill birds, fish, and shellfish, and foul beaches and coastal lands. Also, the Netherlands and Germany are now burning toxic waste at sea. Many people are worried that accidents might add even more poisonous chemicals to the water.

Environmentalists would like to see ocean dumping stopped. They also believe there should be stricter safety regulations for ships that transport dangerous cargo. In the meantime, scientists are studying ways to clean up ocean waters. One researcher, Ananda Chakrabarty, has discovered bacteria that eat oil. He hopes that millions of these tiny creatures might someday digest oil spills.

CHAPTER SIX

NUCLEAR DANGERS

A few years ago, a movie called *The China Syndrome* told the story of a near accident at a *nuclear power* plant. The movie showed scientists working frantically to avoid a nuclear accident. At one point, an actor said that an accident in the plant could wipe out an area the size of Pennsylvania. For several days in March 1979, audiences took that line very seriously. They knew it might come true. At that very moment, scientists were frantically trying to prevent a *meltdown* in a nuclear power plant in Pennsylvania.

Nuclear power plants make electricity, just as coal- and oil-powered plants do. However, nuclear power plants produce energy by splitting atoms, the incredibly tiny particles that make up every bit of matter in the universe. If it is not used properly, nuclear fuel can be very dangerous. The fuel is *radioactive*; it gives off invisible rays that cause cancer and other fatal diseases. One nuclear fuel, *plutonium*, has been called the most dangerous substance on earth. *One thirty-millionth of an ounce* is enough to cause cancer; a lump of plutonium smaller than a softball could kill every person on earth. If air, water, or soil is touched by radiation from plutonium or other radioactive materials, human health can be threatened for hundreds of thousands of years. Therefore every nuclear power plant is built with many safety systems, especially in the *reactor*, the most radioactive part of the plant.

The trouble in Pennsylvania started in a power plant located on Three Mile Island, near the city of Harrisburg. The power plant had begun operating in December 1978. However, it was clear from the beginning that there were many problems with its construction and design. Several times during the first three months of 1979, equipment failed and the plant had to be shut down. Then in March there was a more serious accident, and a bubble of gas formed inside one of the nuclear reactors. Scientists were afraid the bubble would explode, damaging the cooling system of the plant. That's the worst kind of accident possible. Without cooling water, temperatures inside a reactor can climb as high as 4000° F. (2204° C.). The reactor might melt down past the concrete floor of the building and into the earth. If that happened, deadly radiation would be released into the air, endangering thousands of people in the surrounding area, and making much of the state unliveable.

As engineers tried to repair the damage, the governor of Pennsylvania announced that pregnant women, and children under five years of age, should not go within five miles of the plant. (Unborn babies and young children are more easily harmed by radiation than older people.) Pennsylvania officials drew up plans to move a million people out of the area quickly. Luckily, this was not necessary. Though some radioactive gas was released into the air, the reactor was brought under control.

DANGER EVERY STEP OF THE WAY

Accidents at power plants are not the only nuclear dangers. The manufacture of nuclear or "atomic" weapons also creates risks. Nuclear fuel, such as uranium, must be mined from the earth.

The danger of improperly constructed nuclear power plants was brought home to the American public in 1978, when an accident at Three Mile Island near Harrisburg, Pennsylvania, almost lead to the evacuation of local residents.

This is a spent-fuel storage pool for uranium dioxide. After three to five years of use in a nuclear power reactor, the spent fuel is thousands of times more radioactive than it was before it went into the reactor.

According to the United States Public Health Service, an unusually large number of uranium miners die from cancer, perhaps because they have handled radioactive material.

The waste products of uranium mining—huge piles of gravel and dirt called *tailings*—are also unsafe. They remain radioactive for up to a quarter of a million years. Mining companies used to leave tailings unprotected; before their dangers were fully understood, tailings were even used to make bricks. Now tailings are kept inside strong dams. However, accidents still happen. In July 1979, a dam broke near Church Rock, New Mexico, dumping 1100 tons (1001 mt) of radioactive dirt and 94,000,000 gallons (36,000,000 dl) of radioactive water into the Rio Puerco. It was the largest known tailings accident in the United States; there have been fifteen others in the past twenty years. There are now about 140,000,000 tons (127.4 mil mt) of tailings in the United States. By the year 2000, there will be about a billion tons (907,200,000 mt) of the dangerous material.

Waste products from the power and weapons plants are also a worry. Since they are highly radioactive, nuclear wastes must be kept out of the environment and away from people until they lose their radiation. Some radioactive materials are unsafe for "only" 600 years. Other wastes must be safely stored for as long as a half a million years—longer than the entire recorded history of the human race.

At present, nuclear wastes are kept in tanks. No one is satisfied with this method, since there is no container guaranteed not to leak, decay, or open for more than twenty years. Several leaks have already occurred. In 1973, for example, over 100,000 gallons (37,850 dl) of radioactive wastes spilled from a tank in Hanford, Washington.

Scientists are now studying new ways to dispose of radioactive wastes. Salt mines are one possibility, because they tend to stay unchanged for millions of years. Other possible places include the Antarctic ice cap, outer space, and deep under the ocean floor. In 1982, the United States government set up a timetable for a solution to the nuclear waste problem. By 1989, a safe place for the burial of nuclear wastes is to be selected. The storage area is to be ready by 1998. However, no one is sure if these deadlines can be met, because no one yet knows if any place will be safe for the necessary amount of time. What happens during the next ice ages, for example, 10,000 years from now? Can any spot be guaranteed against earthquake for a hundred thousand or more years?

Even if the waste problem is solved, nuclear fuel must still be transported from mines to processing factories, and from factories to power or weapons plants. Each time it is moved, there is the risk that an accident will expose the radioactive material to the environment. In 1976, for example, a truck crashed and spilled 1000 pounds (454 kg) of uranium. Environmentalists also worry about the theft of radioactive material. Criminals or terrorists who succeeded in stealing plutonium or other radioactive material could do tremendous damage.

NUCLEAR BENEFITS

In the face of such danger, it seems strange that nuclear energy is used in the United States at all. Yet many people believe nuclear weapons are necessary for our country's defense. Those who favor nuclear power also point out that it has many advantages. When nuclear power plants are run properly, they do not release sulphur or nitrogen oxides into the air, as coal-burning plants do. Some people also see nuclear plants as a solution to the world's energy needs. One half-inch pellet of uranium can provide as much power as 12 barrels of oil or 3 tons (2.7 mt) of coal. Nuclear power makes America less dependent on expensive, imported oil. Also, when the supply of gas and oil can no longer fulfill the world's energy needs—as early as the year 2000, according to some experts—nuclear power might fill the gap. Though supplies of uranium may run out even sooner, scientists are now working on other types of nuclear plants, such as a "breeder" plant that runs on plutonium and produces more fuel than it uses.

A DIFFICULT DECISION

At the present time about 13 percent of the nation's electricity is produced in nuclear power plants. Still more radioactive material is used by the military to make weapons. According to some people, all these nuclear power plants and "nuke" weapons should be completely done away with. These people believe the benefits of nuclear energy are not worth the risk of a major accident. However, not everyone agrees, and it is likely that the conflict between the two groups will continue for some time. The outcome may very well shape the history of the world.

CHAPTER SEVEN

WHY WE STILL HAVE POLLUTION

No one intended to pollute the environment. No one is in favor of dirty water or dirty air. Yet over the past hundred years, the earth has come to a state of crisis, and in spite of all our cleanup efforts, the pollution is still going on.

This is partly the result of the way we live. When human beings were hunters and gatherers eating animals and wild plants, they had almost no effect on the earth, mainly because the human population was small. Neither did the early, primitive farmers. Yet now we have factories producing hundreds of millions of things for us; our farms use machines and chemicals to grow our food.

Because of modern inventions, our lives are longer. An American baby born in 1900 could expect to live less than fifty years, but today's newborn has a life expectancy of over seventy years. Modern inventions have also made our lives more pleasant. When people talk about the "good old days," they don't mention ice boxes that allowed their food to spoil, or summers without electric fans and air conditioners. Not many Americans would be willing to give up cars, modern plumbing, and plastics, even if that would help to lower pollution.

Although we can't go back, we can change some of our habits. America has less than 6 percent of the world population, but it uses 30 percent of the planet's energy each year. In 1975, the United States wasted more fuel than was used by two-thirds of all the people in the world. Our record since 1975 has been better, mainly because fuel prices have risen sharply in recent years. However, we still waste about half of all the water we use for irrigation, and throw out countless tons of garbage each day that could be recycled. In terms of pollution, an average American family with three children has the same effect on the environment as fifty poor Indian families of the same size.

A great deal of pollution is also the result of ignorance. There are now about five million chemicals known to science. About 300,000 new chemicals are discovered each year. Testing each of those for safety is almost an impossible job. Even identifying the chemicals is often difficult. For example, until recently there were just two laboratories in the United States that were able to test for the presence of dioxin. Today there are only eleven. If dioxin can't be identified, it can't be controlled.

We know very little about how the different parts of our environment are connected. Some of our pollution-control efforts even cause more pollution. The Clean Air Act of 1972, for example, strictly regulated the amount of sulphur and nitrogen dioxides that could be present in the air near each factory. The quickest and cheapest way to lower the amount of pollution was to build taller smokestacks. The pollutants then drifted into high wind currents and were carried far away. In 1970 there were only two smokestacks over 500 feet (1500 dm) high in the United States. By 1980, there were 180, and the air in our cities became cleaner. However, there is only one earth. No pollutant ever really

Pollutants in the air have eaten the marble from the statue's pedestal.

goes away; it just goes somewhere else. In this case, the sulphur and nitrogen oxides mixed with water and became acid rain, one of the worst pollution problems we face today.

It's also hard to know what kind of laws are necessary to insure a clean environment, and how to enforce those laws properly. Since 1970, Congress has passed thirty-one anti-pollution laws. The Environmental Protection Agency is supposed to see that the rules are obeyed. However, there are thousands and thousands of businesses in this country; the EPA simply can't check on every one as often as it should. When the agency studied some of its water-pollution-control stations in 1983, it found that only a third of them were doing an adequate job. The rest didn't have enough people and money to do the work.

There are also people who feel that the present environmental laws are too strict. Los Angeles, for example, has made tremendous efforts to clean up its air, though it still has a long way to go. To meet EPA standards, the city would have to get rid of 90 percent of its cars and 90 percent of its factories. Neither action would be good for Los Angeles, no matter how clean its air became afterwards. Many businesspeople also believe that the environmental laws hold back development unnecessarily. When the Hampton Roads Energy Company wanted to build an oil refinery in Portsmouth, Virginia, the EPA held up approval of the plant for five years. During that period, the cost of the plant jumped from $300,000,000 to $850,000,000.

Pollution control is expensive in other ways. If a factory needs a $250,000 machine to clean its waste water, the cost will be passed on to the people who buy its product in the form of higher prices. Americans also pay higher taxes in order to clean up the environment. During the 1970s, the government and private industry spent $450,000,000,000 on pollution control. During the 1980s, about $690,000,000,000 more will be needed.

Another price for pollution control might be jobs. If a factory owner only has $100,000 to spend and $200,000 is needed to clean its smoke, he or she may have to choose between polluting the air or closing the plant completely. This decision faced the citizens of Tacoma, Washington, in 1983 when the government ruled that the amount of poisonous arsenic in the smoke from a copper plant was too high. According to the Environmental Protection Agency, an estimated four people in Tacoma die each year from cancer caused by the arsenic. The company agreed to spend millions of dollars to lower the amount of arsenic in the air. However, even the lower level could cause one cancer death a year. There seemed to be no way to continue to operate the plant, which employed 600 people, and provide a completely safe environment. Tacoma was forced to choose between the risk of cancer for one of its citizens, and the suffering that unemployment for 600 people would cause. In the country as a whole, about 20,000 jobs were lost between 1972 and 1976 because factories couldn't meet air and water pollution standards. However, in the same period, 1,000,000 new jobs were created in the pollution-control industry. It's also important to remember that pollution itself has an enormous cost.

In spite of the fact that jobs might be endangered and prices might rise, Americans are still in favor of a cleaner environment. In 1981, a poll found that 78 percent of the people in the United States believed factories should be shut down rather than be excused from Clean Air laws. In 1982, 65 percent said that the best water-pollution-control devices available should be used even if that meant losing some jobs. A majority of Americans have also said that they would be willing to pay more for products if the environment were protected.

WHAT YOU CAN DO

Be careful with chemicals. Before you spray the bugs that are eating your garden, find out if there is a natural way to get rid of the pest. Check to see if your clothes and dishes are washed in low-phosphate detergent. If they aren't, see if your household can switch to a different brand. Use "pump" products instead of spray cans. Don't use chemical cleaners if they are not necessary. Sometimes plain water and lots of elbow grease can do the job!

Conserve energy. Don't ride when you can walk; use your bike or public transportation instead of asking for a lift. Turn off lights when they are not needed, and use appliances as little as possible. In spring and fall, adjust your clothing instead of turning on the heat or air conditioning.

Conserve water. Fill the basin instead of letting water run. Keep a jug of ice water in the refrigerator instead of waiting for the faucet to run cold. Use leftover cooking water for plants. Take short showers.

Don't be a litterbug. Put your trash in the proper place, and convince your friends to do the same. Check your community for litter, and see if your class, club, or church group can organize a cleanup project.

Live lightly on the earth. Every product that is manufactured uses water, natural resources, and energy. It eventually becomes garbage, another environmental problem. Before you go shopping, decide what you really need. Recycle old clothes, books, and toys instead of throwing them out. Bring newspapers, cans, and bottles to neighborhood recycling centers.

GLOSSARY

Algae: Tiny plants that live in water. Algae can clog lakes and streams if their growth is unchecked.

Carbon Monoxide: A poisonous gas that is released into the air by automobiles and some factories. In small amounts, carbon monoxide causes headache and nausea. In large amounts it can kill.

Dioxin: A poison that is produced during the manufacture of other chemicals, such as the pesticide 2,4,5-T.

Endrin: A pesticide that is now banned because it can severely damage wildlife.

Ecosystem: A specialized community, including all the component organisms, that forms an interacting system.

Environment: Our physical surroundings: the air, water, soil, and wildlife of the planet.

Environmentalist: A person who is concerned with the well-being of the environment.

Fluorocarbon: A gas used in spray cans.

Heptachlor: A pesticide used on Hawaiian pineapples and on other crops.

Hydrocarbon: A gas released into the air by cars and some factories.

Landfill: Low-lying or swampy land where garbage is dumped.

Leachate: Liquid that seeps from a landfill or toxic waste dump.

Meltdown: An accident in a nuclear power plant in which the reactor overheats.

Nitrogen oxide: A gas released into the air by cars and some factories. When combined with water it can form nitric acid, a part of acid rain.

Nuclear power: Power that is produced by splitting atoms.

Ozone: A gas that is a natural part of our air. In large amounts, ozone is a pollutant.

PCB: A cancer-causing chemical (polychlorinated biphenyl) that was once used in the manufacture of ink, paint, plastic, and electrical parts.

Pesticide: A substance used to kill pests such as insects, rats, and mice, and nuisance plants like weeds and poison ivy.

Phosphate: A substance that is used in detergents and fertilizers.

Plutonium: An extremely radioactive material.

Pollutant: Any substance that dirties the environment.

Polychlorinated Biphenyl: The chemical name of PCB (see above).

Radioactive: Giving off dangerous rays.

Reactor: The part of a nuclear power plant where atoms are split. It is highly radioactive.

Smog: Polluted air that is caused by a combination of gases, usually resulting from automobile exhausts or factory smoke.

Sulphur dioxide: A gas released into the air by factories and cars. When combined with water it can form sulphuric acid, which is part of acid rain.

Tailings: Dangerous, radioactive waste from uranium mining.

Toxic Waste: Poisonous substances left over from the manufacture of other goods.

BOOKS FOR FURTHER READING

Branley, Franklin M. *Water for the World.* New York: Crowell, 1982.

Goode, Stephen. *The Nuclear Energy Controversy.* New York: Franklin Watts, 1980.

Pringle, Laurence. *Nuclear Power.* New York: Macmillan, 1979.

Pringle, Laurence. *Water: The Next Great Resource Battle.* New York: Macmillan, 1982.

Stwertka, Albert and Eve. *Industrial Pollution.* New York: Franklin Watts, 1981.

Sweeney, Robert E. *Environmental Concerns.* New York: Harcourt, Brace, Jovanovich, 1977.

INDEX

Italicized page numbers indicate photographs.